KENTUCKY DM

2019 Permit Practice Test With Over 200 Questions & Answers for Kentucky DMV written Exam

Patience C Anderson

All Rights Reserved. No Contents of this book may be copied, printed or reproduced in any way or by any means without a written consent of the publisher, with exception of brief excerpt in critical reviews and articles.

(General Questions)

1. A full 24 hours of being awake causes impairment that is nearly equal to that of an alcohol content of what?

    a) .02

    b) .05

    c) .08

    d) .10

    Answer: D

2. When you see a vehicle coming toward you in your lane you should turn to the left.

    a) True

    b) False

    Answer: B

3. When you merge with traffic, at what speed should you try to enter traffic?

    a) A slower speed than traffic.

    b) The same speed as traffic.

c) A faster speed than traffic.

d) As fast as you can go.

e) None of the above.

Answer: B

4. **Can you ever be issued a ticket for driving too slowly?**

    a) True

    b) False

    Answer: A

5. **Besides helping you see at night, headlights help other road users see you at any time.**

    a) True

    b) False

    Answer: A

6. **Which of the following emotions would have a great effect on your ability to drive safely?**

    a) Worried

- b) Excited
- c) Afraid
- d) Angry
- e) Depressed
- f) All of the above
- g) None of the above

   Answer: F

7. **Though, both your judgment and vision are affected after drinking alcohol. Which is affected first?**
   - a) Judgment
   - b) Vision

   Answer: A

8. **What color of lines divide lanes of traffic going in opposite directions?**
   - a) White
   - b) Yellow
   - c) Orange

d) Red

e) None of the above.

Answer: B

9. In which of the following situations should one use horn?

a) To notify a vehicle to get out of your way.

b) To warning bike riders that you are passing.

c) When changing lanes quickly.

d) To prevent a possible accident.

e) To let someone know you are angry.

f) All of the above.

Answer: D

10. Below is the proper hand-placement on a steering wheel.

a) True

b) False

Answer: B

Road Signs

(47 Questions on Road Signs)

1. **Put the above signs in their proper order from left to right.**

    a) Guide, warning, stop.

    b) Regulatory, service, stop.

    c) Regulatory, guide, stop.

    d) Warning, service, stop.

    e) Warning, guide, stop.

Answer: E

2. Which sign above is not a route sign?

   a) A.

   b) B.

   c) C.

   d) None of the above.

Answer: C

3. Which of the above sign is found on the back of a slow moving vehicle?

   a) A.

   b) B.

   c) C.

   d) None of the above.

Answer: B

4. **Which sign above is a construction sign?**

    a) A.

    b) B.

    c) C.

    d) None of the above.

    Answer:D

5. **This sign warns that you should**

    a) Hurry and pass someone.

    b) Pass only in an emergency.

    c) Never pass.

    d) Pass on the left.

    e) None of the above.

Answer: C

6. This sign means which of the following?

   a) Look to your left.

   b) Stop if turning left.

   c) Curve to the left.

   d) Sharp turn to the left.

   e) None of the above.

   Answer: D

7. does this sign mean?

   a) 4 way stop ahead.

   b) An intersection of roads ahead.

c) Divided roadway.

d) Yield ahead.

e) None of the above.

Answer: B

8. **Which sign above means do not enter?**

    a) A.

    b) B.

    c) C.

    d) None of the above.

    Answer: B

9. **Which of the above signs mean a divided highway?**

    a) A.

b) B.

c) C.

d) None of the above.

Answer: A

10. It is ok to go 5 mph above most speed limit signs

   a) True

   b) False

   Answer: B

11. What does this sign mean

a) No turning while the light is red even if the road is clear.

b) No turning on green or red.

c) You must make sure the road is clear if the light is red, then you can turn.

d) None of the above.

Answer: A

12. What is the meaning of this sign?

a) Burning building ahead.

b) Firetruck could be entering the road ahead.

c) Truck blocking the road.

d) No trucks ahead.

e) None of the above.

Answer: B

13. This sign means which of the following

   a) Runaway trucks.

   b) Hill.

   c) Truck parked on right triangle ahead.

   d) Truck crossing.

   e) None of the above.

   Answer: B

14. Which sign above means passing permitted

   a) A.

   b) B.

   c) C.

d) None of the above.

Answer: D

15. Which of the following best describes this signs meaning?

 a) Construction worker.

 b) Pedestrian crossing.

 c) School crossing.

 d) Jogging path.

 e) None of the above.

Answer: B

16. This sign means which of the following?

a) Look both ways.

b) Entering 2 way traffic ahead.

c) You can go forward or reverse.

d) One way traffic.

e) None of the above.

Answer: B

17. This sign means which of the following?

a) Multiple curves ahead.

b) Oil on the road.

c) Slippery when wet.

d) Pavement ends.

e) No drunk drivers.

Answer: C

18. This sign means which of the following

    a) Stop.

    b) Stop if going straight.

    c) Stop ahead.

    d) No stop if proceeding straight.

    e) None of the above.

    Answer: C

   A   **B**   **C**

19. What color lettering will be found on Sign A above

a) Red.

b) Black.

c) White.

d) Yellow.

e) None of the above.

Answer: C

20. Which of the above sign is found on the back of a slow moving vehicle?

a) A.

b) B.

c) C.

d) None of the above.

Answer: D

21. This is a real traffic sign

a) True

b) False

Answer: A

22. Which sign above means no passing permitted

a) A.

b) B.

c) C.

d) None of the above.

Answer: C

23. Which sign above is not a guide sign?

a) A.

b) B.

c) C.

d) B and C.

e) A and C.

f) None of the above.

Answer: D

24. This sign shares the same meaning with a _____ sign

a) Stop.

b) Yield.

c) Merge.

d) Stop ahead.

e) None of the above.

Answer: B

A   **B**   **C**

25. Which sign above is a warning sign

    a) A.

    b) B.

    c) C.

    d) None of the above.

    Answer: D

26. What does this orange sign mean?

    a) One handed man with lunch.

    b) Beware of car jacking.

    c) Construction flagger ahead.

    d) Watch for pedestrians.

    e) None of the above.

Answer: C

27. Which sign above is a guide sign?

a) A.

b) B.

c) C.

d) None of the above.

Answer: C

28. Which of the following best describes this signs meaning?

a) Road spins.

b) Drive in circles.

c) Roundabout ahead.

d) Circular highway.

e) None of the above.

Answer: C

29. What does this sign mean?

   a) Road splits

   b) Yield

   c) Merge

   d) Divided highway

   e) Right lane ends

   Answer: D

30. What does this sign mean?

a) Go, yield or stop

b) Yield

c) Stop ahead

d) Signal ahead

e) Caution, no signal at crossing

Answer: D

31. This traffic sign informs you of what?

a) Eat if you are hungry.

b) No silverware beyond this point.

c) Food available at the next exit.

d) his is just highway art.

Answer: C

32. What kind of sign is this?

a) Warning sign.

b) Regulatory sign.

c) Construction sign.

d) Guide sign.

e) None of the above.

Answer: B

33. Yellow signs with black lettering like the one above are called regulatory signs.

a. True

b. False

Answer: B

34. Sign B is a speed limit sign

a) True

b) False

Answer: B

35. What does this sign mean?

a) Look to your right.

b) Stop if turning right.

c) Curve to the right.

d) Sharp turn to the right.

e) None of the above

Answer: C

36. Put the signs above in proper order from left to right.

a) Construction, school crossing, yield.

b) Pedestrian crossing, construction, no passing.

c) School crossing, construction, yield.

d) School crossing, construction, no passing.

e) Pedestrian crossing, construction, yield.

Answer: D

37. What does this sign mean

    a) Construction worker.

    b) Pedestrian crossing.

    c) School crossing.

    d) Jogging path.

    e) None of the above.

    Answer: C

38. What in your opinion is the meaning of the above sign?

    a) This is not a real sign.

b) Pass with care.

c) Uneven shoulder ahead.

d) Narrow bridge ahead.

e) None of the above.

Answer: D

39. Which of the following options best describes this sign?

a) Road splits

b) Yield

c) Merge

d) Divided highway

e) Right lane ends

Answer: C

40. What does this bicycle sign mean?

a) Bicycles race starts here.

b) No bicycles.

c) Bicycle crossing.

d) Bicycles must park by the arrow.

e) None of the above.

Answer: C

41. Sign A above is a state highway sign

a) True

b) False

Answer: A

42. This sign means which of the following

   a) Road splits

   b) Yield

   c) Merge

   d) Divided highway

   e) Right lane ends

   Answer: E

A   B   C

43. Put the signs above in proper order from left to right?

   a) Slow moving vehicle, yield, school zone.

b) Yield, no passing, slow moving vehicle.

c) No passing, slow moving vehicle, yield.

d) School crossing, construction, yield.

e) Pedestrian crossing, slow moving vehicle, yield.

Answer: C

44. Which sign above means two-way-traffic?

a) A.

b) B.

c) C.

d) None of the above.

Answer: C

45. All 3 signs above mean yield

a) True

b) False

Answer: A

46. Square or rectangular signs with blue and white letters or symbols are what kind of signs?

a) Destination

b) Service

c) Route

d) Reference

e) Regulatory

f) Warning

Answer: B

47. Which of the above sign is found on the back of a slow moving vehicle?

e) A.

DEFENSIVE DRIVING TEST.

(28 Question)

1. **Would you drive below the speed limit in poor driving conditions?**

    a) Yes

    b) No

    Answer: A

2. **Would you provide a safe cushion of space when coming back over after a lane change**

a) Yes

b) No

Answer: A

3. Will you only use your horn when necessary?

    a) Yes

    b) No

    Answer: A

4. Will you always avoid using your headlights in an unruly manner?

    a) Yes

    b) No

    Answer: A

5. Will you always pull over to deal with all distractions like phone calls or map reading?

    a) Yes

    b) No

    Answer: A

6. Would you make effort to avoid blocking the passing lanes?

    a) Yes

    b) No

    Answer: A

7. Will you always keep to the right as much as possible?

    a) Yes

    b) No

    Answer: A

8. Would you avoid playing loud music while driving?

    a) Yes

    b) No

    Answer: A

9. Would you use your turn signals every time you turn or change lanes?

    a) Yes

b) No

   Answer: A

10. Do you plan covering the brake when you identify hazards reducing your reaction time?

    a) Yes
    b) No

       Answer: A

11. Would you keep your eyes moving continuously and consciously scanning the road ahead of you as well as your mirrors for hazards?

    a) Yes
    b) No

       Answer: A

12. Would you use your high-beam headlights whenever possible at night?

    a) Yes

b) No

   Answer: A

13. Would you make it a habit to come to a complete stop at all stop signs

    a) Yes

    b) No

    Answer: A

14. Do you plan to always drive with the flow of traffic?

    a) Yes

    b) No

    Answer: A

15. Would you avoid returning inappropriate gestures to other drivers in all situations

    a) Yes

    b) No

    Answer: A

16. Will you always try to yield to faster traffic by moving to the right?

    a) Yes

    b) No

    Answer: A

17. Would you slow down for construction zones and watch out for workers?

    a) Yes

    b) No

    Answer: A

18. Would you avoid driving in the blind spots of other cars and trucks

    a) Yes

    b) No

    Answer: A

19. Would you attempt to get out of the way and avoid confrontation with aggressive drivers

    a) Yes

b) No

Answer: A

20. Will you always pass other vehicles using the left lanes only

    a) Yes

    b) No

    Answer: A

21. Will you always maintain a large following distance at all times?

    a) Yes

    b) No

    Answer: A

22. Will you always use your headlights in all low visibility conditions

    a) Yes

    b) No

    Answer: A

23. Will you follow the right-of-way rules at every 4 way stop signs and intersection?

    a) Yes

    b) No

    Answer: A

24. Will you always avoid talking on the phone and texting while driving?

    a) Yes

    b) No

    Answer: A

25. Will you plan ahead by taking less busy routes at very busy traffic times?

    a) Yes

    b) No

    Answer: A

26. Will you make efforts to be in the proper lane ahead of time so as to avoid last minute lane change?

a) Yes

b) No

Answer: A

27. When you are tired will you avoid driving?

    a) Yes

    b) No

    Answer: A

28. Will you always make sure that you do not ever put yourself in a situation where you end up drinking and driving?

    a) Yes

    b) No

    Answer: A

PRACTICE TEST

(66 Questions on Practice Test 1)

1. What do white painted curbs indicate?

    a) Loading zone for freight or passengers.

b) No loading zone.

c) Loading zone for passengers or mail.

d) No loitering.

e) None of the above.

Answer: C

2. Name some places where you may find slippery spots on the road.

a) In corners and at stop signs.

b) Shady spots and on overpasses and bridges.

c) In tunnels and on hills.

d) Near large bodies of water.

e) None of the above.

Answer: B

3. Night driving is very dangerous because?

a) Traffic signs are less visible at night.

b) The distance we can see ahead is reduced.

c) People are sleepy at night.

d) Criminals come out at nighttime.

e) None of the above.

Answer: B

4. **Which of the following would be most effective in avoiding a collision?**

    a) Keeping your lights on at all times.

    b) Wearing a seat belt.

    c) Driving in daytime hours only.

    d) Keeping a cushion of space at all times.

    e) Driving slow at all times

    Answer: D

5. **What drugs would affect your ability to drive safely?**

    a) Almost any drug prescription or even over the counter drugs, can affect your ability to drive.

    b) Alcohol and marijuana.

    c) Only illegal drugs.

    d) None of the above.

    Answer: A

6. **In driving a roundabout, the same general rules apply as for maneuvering through any other type of intersection.**

    a) True

    b) False

    Answer: A

7. **When a traffic signal light turns green, what would you do?**

    a) Accelerate as quickly as possible.

    b) Yield to pedestrians.

    c) Count two seconds before accelerating.

    d) Do not move until another driver signals for you to go.

    e) None of the above.

    Answer: B

8. **At an intersection with a flashing red light, what should you do?**

a) Firstly, come to a full stop thereafter, go when it is safe to do so.

b) Come to a full stop, then go when it flashes green.

c) Slow down and yield to any vehicles already in the intersection.

d) Come to a full stop, then go when it turns solid green.

Answer: A

9. When more than one vehicle arrives at the same time at a four way stop, which vehicle should go first?

a) The first one that attempts to go.

b) The vehicle on the left.

c) The vehicle on the right.

d) None of the above.

Answer: C

10. When is it permissible for a driver to use a handicapped parking space?

    a) Only if a physically handicapped person is in the motor vehicle when it is parked.
    b) Only if a physically handicapped person is being delivered or picked up.
    c) Only if the vehicle displays a handicapped person placard or license plates.
    d) All of the above.
    e) None of the above.

    Answer: C

11. Scanning and seeing events very well in advance would really help to prevent what?

    a) Fatigue
    b) Distractions
    c) Panic stops
    d) Lane changes
    e) None of the above

Answer: C

12. A white square or a rectangular signs with white, red, or black letters or symbols are usually what kind of signs?

   a) Destination

   b) Service

   c) Route

   d) Reference

   e) Regulatory

   f) Warning

   Answer: E

13. When is it appropriate to obey instructions from school crossing guards?

   a) Only during school hours.

   b) Only if you see children present.

   c) Only if they are licensed crossing guards.

   d) Only at a marked school crosswalk.

e) At all times.

Answer: E

14. **Which way should you turn your wheels if any, when packing on a downgrade with a curb?**

    a) Turn them away from the curb.

    b) Point them straight ahead.

    c) Turn them toward the curb.

    d) Park facing up the hill instead.

    e) None of the above.

    Answer: C

15. **Talking on a cell phone may increase your chances of being in a crash by as much as four times.**

    a) True

    b) False

    Answer: A

16. **If a tire blows out what should you do?**

a) Hold the steering wheel firmly while easing up on the gas pedal.
b) Apply the brakes firmly.
c) Shift to neutral and apply the brakes.
d) Speed up to gain stability first, then pull over.
e) None of the above.

Answer: A

17. When you have a green light, but traffic is backed up into the intersection, what should you do?

a) Enter the intersection immediately and hope traffic clears before the light changes.
b) Wait until traffic clears before entering the intersection.
c) Try to go around the traffic.
d) Sound your horn to clear the intersection.
e) None of the above.

Answer: B

18. **Where more than one vehicle is stopped at an intersection, which vehicle has the right-of-way?**

   a) The largest vehicle.
   b) The first vehicle that attempts to go.
   c) The first vehicle to arrive.
   d) The vehicle on the right.
   e) None of the above.

   Answer: C

19. **What is the acceptable minimum safe following distance under most conditions?**

   a) A minimum of 2 seconds is the recommended following distance under most conditions.
   b) The minimum recommended following distance under most conditions is 4 seconds.
   c) A minimum of 5 seconds is the recommended following distance under most conditions.

d) The minimum recommended following distance under most conditions is 6 seconds.

e) None of the above.

Answer: B

20. **If you are at a highway entrance and has to wait for a gap in traffic before entering the roadway, what should you do?**

    a) Pull up as far as you can on the ramp and wait leaving room behind you on the ramp for other vehicles.

    b) Drive to the shoulder and wait for a gap in the roadway, then accelerate quickly.

    c) Slow down on the entrance ramp to wait for a gap, then speed up so you enter at the same speed that traffic is moving.

    d) Slow down on the entrance ramp to wait for a gap, then sound your horn and activate your emergency flashing lights to alert drivers you are entering the roadway.

e) None of the above.

Answer: C

21. When is it really ok to drive faster than the posted speed limit?

a) When being followed too closely.

b) To keep pace with the flow of traffic.

c) If you have an emergency.

d) It is never ok.

Answer: D

22. You can cross a solid yellow line to do which of the following?

a) Making a U turn on the highway.

b) To turn into a driveway if it is safe to do so.

c) To pass on the highway.

d) To get a better view of the road ahead.

Answer: B

23. How would you see if there is a vehicle in your blind spot?

   a) Lean back and forth looking in your mirrors.

   b) Look over your shoulder.

   c) Adjust your car's power mirrors if you have them.

   d) Nothing can be done, that is why it is called a blind spot.

   Answer: B

24. On a rainy, a snowy, or foggy days when it begins to get dark, and when driving away from a rising or setting sun, it is a good time to?

   a) Check the tires.

   b) Put on your seatbelt.

   c) Turn on your headlights.

   d) Roll up the windows.

   e) None of the above.

   Answer: C

25. **Name some likely places where you may find slippery spots on the road.**

    a) In corners and at stop signs.

    b) Shady spots on overpasses and bridges.

    c) In tunnels and on hills.

    d) Near large bodies of water.

    e) None of the above.

    Answer: B

26. **If you want to turn left at an intersection but the oncoming traffic is heavy, what would you do?**

    a) Wait at the crosswalk for traffic to clear.

    b) Wait in the center of the intersection patiently for the traffic to clear.

    c) Start your turn forcing others to stop.

    d) None of the above.

    Answer: B

27. What does a flashing yellow light indicate?

a) Pedestrian crossing.

b) School Crossing.

c) Stop, then proceed with caution.

d) Proceed with caution.

e) None of the above.

Answer: D

28. Whenever driving in heavy fog during the daytime you should drive with your?

a) Headlights off.

b) Parking lights on.

c) Headlights on low beam.

d) Headlights on high beam.

e) None of the above.

Answer: C

29. Which of these statements is correct?

a) A solid or dashed yellow line show that the left edge of traffic lanes going in your direction.

b) A solid or dashed yellow line indicates the right edge of traffic lanes going in your direction.

Answer: A

30. After drinking alcohol, a cold shower or coffee will lower your blood alcohol content.

a) True

b) False

Answer: B

31. Put the following signs in their proper order from left to right.

a) Yield, school zone, construction.

b) School zone, construction, no passing.

c) Pass with care, construction, yield.

d) No passing, slow moving vehicle, yield.

e) No passing, construction, yield.

Answer: D

32. There are certain situations where it is legal to double park.

a) True

b) False

Answer: B

33. Safety belts would help you keep control of your car.

a) True

b) False

Answer: A

34. What would you do if you are at an intersection and you hear a siren?

a) Stop and do not move until the emergency vehicle has passed.

b) Continue through the intersection, then pull over to the right side of the road and stop.

c) Back up and pull over to the right side of the road and stop.

d) None of the above.

Answer: B

35. The state examiner would check the person's vehicle before beginning the driving test to:

a) Be sure that the vehicle has all the required equipment.

b) To be sure that the vehicle is in safe operating condition.

c) Check for cleanliness.

d) A and C.

e) A and B.

Answer: E

36. **Passing is permissible in either direction if there are two solid yellow lines in the center of the road when?**

    a) When following a slow truck.

    b) Only if you are sure it is safe.

    c) On US highways only.

    d) Passing is never permitted.

    Answer: D

37. **A driver turning left at an intersection is expected to yield to what?**

    a) Vehicles approaching from the opposite direction.

    b) Pedestrians, vehicles, and bicycles approaching from the opposite direction.

    c) Pedestrians, vehicles, and bicycles approaching from the right.

    d) Pedestrians, vehicles, and bicycles approaching from the left.

e) None of the above.

Answer: B

38. What does a yellow sign indicate?

a) State highway ahead.

b) A special situation or a hazard ahead.

c) Construction work ahead.

d) Interstate Sign.

e) None of the above.

Answer: B

39. One of the basic things to remember about driving at night or in fog is to?

a) Be ready to brake more quickly.

b) Watch for cars at intersections.

c) Drive within the range of your headlights.

d) Use your high beams at all times.

e) None of the above.

Answer: C

40. When you are driving faster than other traffic on a freeway, which lane should you use?

   a) The right lane.

   b) The shoulder.

   c) The left lane.

   d) The carpool lane.

   e) None of the above.

   Answer: C

41. How far ahead can you look when you are on the highway?

   a) 3 to 5 seconds.

   b) 5 to 10 seconds.

   c) 10 to 15 seconds.

   d) 15 to 20 seconds.

   e) None of the above.

   Answer: C

42. A white square or a rectangular signs with white, red, or black letters or symbols are usually what kind of signs?

    a) Destination
    b) Service
    c) Route
    d) Reference
    e) Regulatory
    f) Warning

    Answer : B

43. When you see a stopped vehicle on the side of the road what should you do?

    a) Stop and offer assistance.
    b) Slow down and use caution when passing.
    c) Sound the horn to let them know you are about to pass them.
    d) Notify emergency services.
    e) None of the above.

Answer: B

44. When is it really illegal to turn right at a red light?

   a) It is always legal.

   b) It is never legal to turn right at a red light.

   c) Only on one-way roads.

   d) Only if there is a no-turn-on-red sign posted.

   e) None of the above.

   Answer: D

45. When you see a car approaching on your lane you should:

   a) Pull to the right and slow down.

   b) Sound your horn.

   c) At night, flash your lights.

   d) All of the above.

   e) None of the above.

   Answer: D

46. It is legal to park next to a fire hydrant as long as you move your vehicle if necessary.

a) True

b) False

Answer: B

47. In which of the following conditions would you need extra following distance?

a) When driving on slippery roads.

b) When following a motorcycle.

c) When following trucks or vehicles pulling trailers.

d) When it is hard to see.

e) None of the above.

f) All of the above.

Answer: F

48. The Yellow diamond signs with black letters or symbols are what kind of signs?

   a) Destination
   b) Service
   c) Route
   d) Reference
   e) Regulatory
   f) Warning

   Answer: F

49. It is legal and acceptable to continue at the same speed when an emergency vehicle approaches you with a siren and flashing lights if there is another lane open.

   a) True
   b) False

   Answer: B

50. You have to yield to traffic on your left already in the roundabout.

    a) True
    b) False

    Answer: A

51. Which of the following about littering while driving is true?

    a) It may cause a traffic accident.
    b) Is against the law.
    c) It can lead to large fines up to and including jail time.
    d) All of the above.
    e) None of the above.

    Answer: D

52. Which of these influences the effects of alcohol in the body?

    a) The time how difference between each drink.

b) The body weight of a person.

c) The amount of food in the stomach.

d) All of the above.

e) None of the above.

Answer: D

53. Both of these signs indicate that you are entering a School Zone.

a) True

b) False

Answer: B

54. Multiple lanes of travelling in the same direction are separated by lane markings of what color?

a) Red

b) Yellow

c) White

d) Orange

Answer: C

55. How many drinks would it take to affect your driving?

   a) 1

   b) 2

   c) 3

   d) 4

   e) 5

   Answer: 1

56. What do an orange sign mean?

   a) State highway ahead.

   b) Merging lanes ahead.

   c) Construction work ahead.

   d) Divided highway ahead.

   e) None of the above.

   Answer: C

57. What would be the correct action to take when at a railroad crossing that does not have signals?

    a) Come to a complete stop.

    b) Slow down and be prepared to stop.

    c) Speed to get across the tracks quickly.

    d) None of the above.

    Answer: B

58. To keep a steady speed and signaling in advance when slowing down or turning will help maintain what

    a) A safe distance ahead of your vehicle.

    b) A safe distance behind your vehicle.

    c) A safe distance next to your vehicle.

    d) All of the above.

    e) None of the above.

    Answer: B

59. Which of the signs below means you are entering a School Zone?

a) A.

b) B.

c) C.

d) None of the above.

Answer: D

60. A solid white line shows what part of the traffic lane on a road?

a) It separates lanes of traffic moving in the opposite direction. Single white lines do also mark the right edge of the pavement.

b) It separates lanes of traffic moving in the same direction. Single white lines do also mark the left edge of the pavement.

c) It separates lanes of traffic moving in the same direction. Single white lines do also mark the right edge of the pavement.

d) It separates lanes of traffic moving in the opposite direction. Single white lines do also mark the left edge of the pavement.

e) None of the above.

Answer: C

61. When you make a left turn from a two-way street to a one-way street in which lane should your vehicle be in when the turn is completed?

a) In the right lane.

b) In the left lane.

c) In the center lane.

d) None of the above.

Answer: C

62. Which of these practices is not only dangerous but illegal to do while driving?

a) Adjusting your outside mirrors manually.

b) Wearing headphones that cover both ears.

c) Putting on make-up.

d) Eating and drinking coffee.

e) Reading a map.

Answer: B

63. If a traffic light changes from green to yellow as you approach an intersection. What should you do?

a) Keep going at your current speed.

b) Stop before the intersection.

c) Stop, even if in the intersection.

d) Speed up to beat the light before it turns red.

e) None of the above.

Answer: B

64. If you plan to pull into a driveway immediately after an intersection. When should you signal?

a) After you cross the intersection.

b) Before you cross the intersection.

c) When you start your turn.

d) In the middle of the intersection.

Answer: D

65. What would you do if a railroad crossing has no warning devices?

a) Stop about 15 feet away from the railroad crossing.

b) With an increased speed, you can cross the tracks quickly.

c) Slow down and then proceed with caution.

d) All crossings have a control device.

e) None of the above.

Answer: C

66. What do white painted curbs indicate?

f) Loading zone for freight or passengers.

g) No loading zone.

h) Loading zone for passengers or mail.

i) No loitering.

j) None of the above.

Answer: C

(55 Questions on Practice Test 2)

1. Which of these places should you never park?

    a) On sidewalks.

    b) In bicycle lanes.

    c) In front of driveways.

    d) By fire hydrants.

    e) All of the above.

    Answer: E

2. What does this hand signal signify?

   a) Left turn.

   b) Right turn.

   c) Stop or slowing down.

   d) Backing.

   e) None of the above.

   Answer: B

3. **If driving in the rain or snow during the day you should?**

   a) Use your high beams.

   b) Use your fog lights.

   c) Use your low beams.

   d) Use no headlights.

   Answer: B

4. Hitting any vehicle moving in the opposite direction is more better than hitting a vehicle moving in the same direction.

   a) True
   b) False

   Answer: B

5. If you have entered an intersection already when the light changes, you should?

   a) Stop in the intersection.
   b) Proceed and clear the intersection.
   c) Flash your lights through the intersection.
   d) Sound your horn through the intersection.

   Answer: B

6. What do blind spots mean?

   a) Blind spots are places for blind people to cross at an intersection.
   b) Blind spots are dots often seen by drivers who have been drinking.

c) Blind spots are such areas near the left and right rear corners of your vehicle that you cannot see in your rear-view mirrors.

   d) Blind spots are spots seen after staring into oncoming headlights at night.

   e) None of the above.

   Answer: C

7. **If you hurry it is ok to pass when approaching the top of a hill or a curve.**

   a) True

   b) False

   Answer: B

8. **Driving very much slower than the speed limit in normal conditions may do what?**

   a) It can decrease the chances of an accident.

   b) It does not change anything.

   c) It can increase driver safety.

   d) It can increase the chance of an accident.

Answer: D

9. It is a must that you use high beam lighting in fog, snow, and heavy rain.

    a) True

    b) False

    Answer: B

10. You must avoid passing on the right if it means driving off the paved or main portion of the roadway.

    a) True

    b) False

    Answer: A

11. If a driver is approaching an intersection, with the traffic light showing green and the driver wants to drive straight through. While another vehicle is already in the intersection making a left turn. Who has the right-of-way?

    a) The driver who wants to drive straight.

b) The driver who is turning left.

   Answer: B

12. A cross-buck sign or a white X shaped that says Railroad Crossing on it has the same meaning as a stop sign.

    a) True

    b) False

    Answer: B

13. Whenever you pass a vehicle traveling in the same direction, you should pass on the left.

    a) True

    b) False

    Answer: A

14. Which lane should you end up in after completing your turn, when turning from one of three turn lanes?

    a) The right lane if clear.

    b) The lane you started in.

c) The left most lane.

d) Always the middle lane.

Answer: B

15. **If you are traveling 47 mph on a highway with a speed limit of 55 mph, which lane should you be in?**

    a) The far left lane.

    b) The carpool lane.

    c) The far right lane.

    d) The middle lane.

    e) The bicycle lane.

    Answer: C

16. **A pedestrian that uses a white or a red-tipped white cane is usually what?**

    a) A policeman.

    b) A construction worker.

    c) A blind person.

    d) A crossing guard.

Answer: C

17. Traffic Light Meaning: The light usually changes from green to red. Be prepared to stop for the red light.

    a) Red arrow.

    b) Steady yellow.

    c) Flashing yellow.

    d) Green arrow.

    e) Steady green.

    f) None of the above.

Answer: B

18. When you experience glare from a vehicles headlights at night you should?

    a) Look above their headlights.

    b) Look below their headlights.

    c) Fix your eyes toward the right edge of your lane.

    d) Look toward the left edge of your lane.

Answer: C

19. What does a red painted curb indicate?

a) Loading zone.

b) Reserved for passenger pick up or drop off.

c) No parking or stopping.

Answer: C

20. What do you think this hand signal mean?

a) Left Turn.

b) Right Turn.

c) Stop or Slowing Down.

d) Backing.

e) None of the above.

Answer: A

21. You can cross a double yellow line to pass another vehicle if only the yellow line next to you is what?

    a) A solid line.

    b) A thinner line.

    c) A broken line.

    d) A thicker line.

    e) None of the above.

    Answer: C

22. After being pulled over by law enforcement, you should immediately exit the vehicle and quickly approach the officer's police car.

    a) True

    b) False

    Answer: B

23. Traffic Light Meaning: Stop, yield to the right-of-way, and go when it is safe.

    a) Red arrow.

b) Steady yellow.

c) Flashing yellow.

d) Green arrow.

e) Flashing red.

f) None of the above.

Answer: E

24. What is the minimum safe-following distance under most conditions?

a) The recommended minimum following distance under most conditions is 2 seconds.

b) The recommended minimum following distance under most conditions is 4 seconds.

c) The recommended minimum following distance under most conditions is 5 seconds..

d) The recommended minimum following distance under most conditions is 6 seconds.

e) An unbalanced load with too much weight on any one axle.

f) None of the above.

Answer: B

25. It is not necessary to signal before pulling away from a curb or exits a freeway.

    a) True

    b) False

    Answer: A

26. While operating a motor vehicle, your both hands should be on the steering wheel at all times except you are texting

    a) True

    b) False

    Answer: B

27. Eating, drinking, angry, ill, and texting while driving are all examples of?

    a) Safe Driving

    b) Defensive Driving

    c) Talented Driving

d) Distracted driving

Answer: D

28. Swerving right instead of toward oncoming traffic to prevent a crash is the best thing to do.

a) True

b) False

Answer: A

29. A Pentagon shaped sign signify which of the following?

a) No Passing Zone.

b) Railroad Crossing.

c) School Zone.

d) Yield.

e) Stop.

Answer: C

30. Unless prohibited by a sign, when can one turn left at a red light?

a) In an emergency.

b) From a two way road to a one way road.

c) From a one way road to a one way road.

d) Never.

Answer: C

31. **Keep your eyes on the road, quickly shift to neutral, pull off the road when it is safe to do so, then turn off the engine are the procedures for which of the following?**

    a) A stuck gas pedal.

    b) A power failure.

    c) Brake failure.

    d) None of the above.

    Answer: A

32. **You should always avoid placing infants or a small child in the front seat of a vehicle with airbags.**

    a) True

    b) False

Answer: A

33. **Passing on the right is permitted when it is safe and the driver of the other vehicle is making a left turn.**

    a) True

    b) False

    Answer: A

34. **When you drive on slippery roads, you should increase your following distance by _____.**

    a) 2 times.

    b) 3 times.

    c) 4 times.

    d) 5 times.

    Answer: A

35. **While operating a motor vehicle, your both hands should be on the steering wheel at all times except you are texting**

    c) True

d) False

Answer: B

36. If driving on a highway posted for 65 mph and the traffic is traveling at 70 mph, how fast would you legally drive?

a) As fast as the speed of traffic.

b) Between 65 mph and 70 mph.

c) As fast as the speed needed to pass other traffic.

d) No faster than 65 mph.

Answer: D

37. In which of these situations is it ok to back up on the highway?

a) If you miss your exit.

b) To go back to see an accident.

  c) To pick up someone on the side of the highway.

  d) It is never ok to back up on the highway.

  Answer: D

38. A driver who approaches an intersection should yield the right-of-way to traffic that is at the intersection.

  a) True

  b) False

  Answer: A

39. When you are on the highway how far ahead should you look?

  a) A block away.

  b) A quarter mile

  c) A half mile

  d) A full mile

  Answer: B

40. Most rear-end collisions are caused by the vehicle at the back following too closely.

    a) True
    b) False

    Answer: A

41. When should you realy obey a construction flagger's instructions?

    a) Only if you see it is necessary to do so.
    b) If they do not conflict with existing signs or signals.
    c) If they are wearing a state badge.
    d) At all times in construction zones.
    e) None of the above.

    Answer: D

42. When another driver follows you too closely you should?

    a) Slowly speed up.
    b) Jam the brakes.

c) Flash your brake lights 3 times.

d) Move to another lane if there is room.

e) None of the above.

Answer: D

43. The picture below shows an improper hand placement on the steering wheel.

a) True

b) False

Answer: A

44. Keeping your eyes always locked straight ahead is a good defensive driving practice.

a) True

b) False

Answer: B

45. Broken yellow lines separate lanes of traffic going in the same direction.

    a) True

    b) False

    Answer: B

46. When exactly should safety belts be worn?

    a) At all times as a driver and as a passenger.

    b) Only when driving on curvy roads.

    c) Only when riding in the back seat.

    d) Only when driving on the freeway.

    e) None of the above.

    Answer: A

47. When you park on an uphill grade, which way should you turn your wheels?

    a) Left

    b) Right

    c) Straight

    d) None of the above

Answer: A

48. When you park on an uphill grade, which way should you turn your wheels?

   e) Left

   f) Right

   g) Straight

   h) None of the above

   Answer: A

49. The best way to enter a freeway smoothly is to accelerate on the entrance ramp to match the speed of freeway traffic on the right lane.

   a) True

   b) False

   Answer: A

50. Pavement line colors normally show if you are on a one-way or two-way roadway.

   a) True

   b) False

Answer: A

51. What does this hand signal indicate?

    a) Left turn.

    b) Right turn.

    c) Stop or slowing down.

    d) Backing.

    e) None of the above.

Answer: C

52. Traffic Light Meaning: You must not go in the direction of the arrow until the light is off and a green light or green arrow light goes on.

    a) Red arrow.

    b) Steady yellow.

    c) Flashing yellow.

d) Green arrow.

e) Steady green.

f) None of the above.

Answer: A

53. When you parking next to a curb, how close should you park when?

a) No closer than 6 inches from the curb.

b) No farther than 6 inches from the curb.

c) Not any closer than 12 inches from the curb.

d) No farther than 18 inches from the curb.

e) No farther than 24 inches from the curb.

Answer: D

54. Approaching an intersection with traffic control signals that are not working, you should treat it as you would a 4-way yield sign.

a) True

b) False

Answer: A

55. Which of these is a good example of defensive driving?

a) Keeping an eye on the cars brake lights in front of you while driving.

b) Keeping your eyes moving to look for possible hazards.

c) Putting one car length between you and the car ahead of you.

d) Checking your car's mirror a couple times per trip.

Answer: B

56. How far ahead should you look when you are driving in town?

a) 1 city block.

b) 1 quarter mile.

c) 1 half mile.

d) 1 mile.

Answer: A

TEEN DRIVER SAFETY

(14 Questions)

1. **Which age group is 3 times more likely to die in a motor vehicle crash than the average of all other drivers combined?**

    a) 16

    b) 17

    c) 18

    d) 19

    Answer: A

2. **The risk of vehicle crashes are higher for which age group over all other groups?**

a) 16 to 19

b) 20 to 22

c) 23 to 25

d) Age does not matter

Answer: A

3. 22 percent of drivers between the age of 15 and 20 involved in fatal crashes were drinking in 2010

    a) True

    b) False

    Answer: A

4. Teens have the lowest rate of seat belt use compared with other age groups.

    a) True

    b) False

    Answer: A

5. Will you share this quiz with a teen you care about?

    a) Yes

    b) No

    Answer: A

6. Two hundred and eighty two thousand teens were injured in vehicle crashes in 2010

a) True

b) False

Answer: A

7. What percent of teenage motor vehicle crash deaths in 2008 were passengers in the vehicles?

   a) 22

   b) 37

   c) 56

   d) 81

   Answer: D

8. What is the leading cause of death for teens in the US?

   a) Cancer

   b) Suicide

   c) Auto crashes

d) Murder

   Answer: C

9. Teenagers between the ages of 16 to 19 are ___ times more likely than drivers 20 and over to be in a fatal crash

   a) 0
   b) 2
   c) 3
   d) 4

   Answer: C

10. 16 to 17 year old driver fatality rates decrease with each additional passenger added to a vehicle.

    a) True
    b) False

    Answer: B

11. 63 percent of teenage passenger deaths in 2008 occurred in vehicles driven by another teenager.

    a) True

b) False

Answer: A

12. **How many high school teens drank alcohol and drove in 2011?**

    a) 1,000 teens.

    b) 10,000 teens.

    c) 100,000 teens.

    d) 1,000,000 teens

    Answer: D

13. **55 percent of teen driving deaths occur on which days?**

    a) Monday and Tuesday

    b) Wednesdays

    c) Thursdays

    d) Friday, Saturday, and Sunday

    Answer: D

14. **The vehicle death rate for teen male drivers and passengers is almost twice that of females.**

a) True
b) False

Answer: A

f) B.
g) C.
h) None of the above.

Answer: B

Made in the USA
Lexington, KY
02 June 2019